FRIENDS
OF THE MISSISSAUGA LIBRARY SYSTEM
Thank you for your
SUPPORT

D1318607

Materials, Materials, Materials

Rock

Chris Oxlade

Heinemann Library
Chicago, Illinois

Published by Heinemann Library,
an imprint of Reed Educational & Professional Publishing,
Chicago, Illinois

Customer Service 888-454-2279

Visit our website at www.heinemannlibrary.com

Designed by Storeybooks
Originated by Ambassador Litho
Printed and bound in Hong kong/China

06 05 04 03 02
10 9 8 7 6 5 4 3 2

Library of Congress Cataloging-in-Publication Data
Oxlade, Chris.
Rock / Chris Oxlade.
p. cm. -- (Materials, materials, materials)
Includes bibliographical references and index.
Summary: Presents an overview of igneous, sedimentary, and metamorphic rocks, and their properties, formation, and uses.
ISBN 1-58810-585-7 (lib. bdg.) ISBN 1-4034-0086-5 (pbk. bdg.)
1. Rocks--Juvenile literature. 2. Stone--Juvenile literature. [1. Rocks 2. Stone.] I. Title.

QE432.2 .O95 2002
552--dc21
2001003926

Acknowledgments
The author and publishers are grateful to the following for permission to reproduce copyright material: p. 4 Andrew Brown/Ecoscene; pp. 5, 29 Martyn Chillmaid; p. 6 Winkley/Ecoscene; p. 7 Dr. B. Booth/GSF Picture Library; pp. 8, 9, 18, 24 GSF Picture Library; pp. 10, 15, 17 Corbis; pp. 11, 23 Tudor Photography; p. 12 Tony Page/Ecoscene; p. 13 Harold Taylor Abipp; p. 14 Ecoscene; p. 16 Barry Webb/Ecoscene; p. 19 Piers Cavendish/Impact; p. 20 David Kampfner; p. 21 Eye Ubiquitous; p. 22 H. Rogers/Tripp; p. 25 Anthony Cooper/Ecoscene; p. 26 Stuart Bebb/Oxford Scientific Films; p. 27 Tony Page/Ecoscene

Cover photograph reproduced with permission of Tudor Photography.

Every effort has been made to contact copyright holders of any material reproduced in this book. Any omissions will be rectified in subsequent printings if notice is given to the publisher.

Some words are shown in bold, **like this.** You can find out what they mean by looking in the glossary.

Contents

What Is Rock?

Rock is a **natural** material. A thick layer of rock covers the earth. It is called the earth's **crust.** It is easy to see this rock in hills and mountains.

Rock is a useful material. We use rock for building homes and roads. We also use it for decorating buildings and making jewelry.

Where Rock Comes From

This rock was made from sand that piled up on the bottom of the sea millions of years ago. The sand was buried and pressed together to make the rock.

This rock is called limestone. It is made from sea creatures. When the creatures died, they fell to the bottom of the sea. Over many years, their skeletons were pressed together to make limestone.

More Types of Rock

The earth's **crust** is made of hard rock. Underneath the crust is hot, runny rock called lava. Sometimes lava leaks through the crust. When it cools down, lava turns into rock.

Over many years, rock becomes buried under more and more rock. Deep underground, heat changes the rock that is buried into a new, different kind of rock.

Hard and Soft

There are many different types of rock. Many rocks are very hard. This kitchen counter is made from a hard rock called granite.

Some rocks are soft. Some are so soft that you can break them with your hands. Chalk is a soft, white rock. It is used for writing.

Color and Pattern

Rocks are made from materials called **minerals.** Minerals can be black, white, pink, green, and other colors. The color of a rock depends on the minerals in it.

Rocks have different patterns of color in them. Some have stripes. Some have swirls. A type of rock called marble is known for its pretty patterns of color.

Weathering

The rock on the earth's surface is worn down by rain and wind. It is also heated by the sun, which makes it break. Slowly, it breaks into small pieces. This is called **weathering.**

Pieces of rock are washed into streams and rivers. They bump along in the water. As they do, they wear away the bottom and sides of the river. Over millions of years, this forms giant **canyons.**

Finding Rock

Rock must be dug out of the ground before it can be used. This happens in places called **quarries.** Sometimes the rock is broken up with **explosives** to make it easier to dig up.

Gravel is made of small, round pieces of rock. We get gravel from places called gravel pits. Huge machines scoop up the gravel and put it into trucks.

Building with Rock

Most of the rock we dig from the ground is used for building. We use hard, strong rock for making the walls of houses and other buildings.

These workers are building with a material called concrete. It is made from **gravel,** sand, and **cement.** When the concrete dries, it becomes very hard and strong.

Rock on the Ground

Hard rock does not wear away easily. It is good for making floors that many people walk over. In hot countries, floors are often made of rock. The rock helps keep the room cool.

When you ride in a car, you are rolling along on rock. Many roads are made from a mixture of bits of hard rock and sticky **tar.**

Rock Decoration

We use rocks with interesting colors and patterns for decoration. This building is covered in a rock called granite. It has been polished to make it shiny.

Shiny pieces of rock are also used to make jewelry. Sometimes the pieces of rock are cut from bigger rocks. Sometimes they are small pebbles.

Making Rock Shapes

Artists called sculptors use rock to make **sculptures.** They start with a block of rock. Then they cut bits off with sharp tools until the rock is the shape they want.

Pieces of rock are also cut into shapes to decorate buildings. The people who cut the rock are called **stonemasons.** They often cut new pieces of rock to fix the old rock on buildings.

Rock and Pollution

Taking rock from the ground can make a mess. And trucks carrying rock from **quarries** make a lot of noise and dirt.

Sometimes, **pollution** from cars and **factories** mixes with rain. When that rain falls on rock, it slowly wears it away. This rock **sculpture** has been spoiled by pollution.

Fact File

- Rock is a **natural** material.
- Some types of rock, such as granite, are very hard.
- Some types of rock, such as chalk, are very soft.
- Rocks come in many different colors.
- Rocks have different patterns of color in them.
- Rock does not burn when it is heated.
- Rock does not let **electricity** flow through it.
- Rock is not attracted by **magnets.**

Can You Believe It?

Pumice is a special type of rock that comes from volcanoes. It is full of air bubbles that make it very light. Pumice is the only rock that floats!

Glossary

canyon deep valley with steep, rocky sides

cement thick, gooey material that becomes hard and strong when it dries

crust thick layer of rock that covers the earth

electricity form of power that can light lamps, heat houses, and make things work

explosive a material that blows up when it is heated

factory big building where things are made using machines

gravel small, round pieces of rock

magnet piece of iron or steel that pulls iron or steel things toward it

mineral natural material that rock is made from

natural comes from plants, animals, or rocks in the earth

pollution harmful chemicals in the air, rivers, and seas

quarry place where rock is dug from the ground

sculpture work of art made from rock or other materials, such as metal or wood

stonemason someone who cuts rock into shapes for buildings

tar gooey, black material that is made from oil

weathering when rock is worn away over time by wind and rain

More Books to Read

Ashwell, Miranda and Owen, Andy. *Mountains*. Chicago: Heinemann Library, 1998.

Royston, Angela. *Materials*. Chicago: Heinemann Library, 2001.

Flanagan, Alice. *Rocks*. Minneapolis: Compass Point Books, 2000.

Russel, William. *Rocks & Minerals*. Vero Beach, Fla.: Rourke, 1994.

Index